THE
SENSES

CHELSEA HOUSE
PUBLISHERS

A Haights Cross Communications Company ®

Philadelphia

First hardcover library edition published
in the United States of America in
2006 by Chelsea House Publishers,
a subsidiary of Haights Cross Communications.
All rights reserved.

A Haights Cross Communications ✈ Company ®

www.chelseahouse.com

Library of Congress Cataloging-in-Publication
applied for.
ISBN 0-7910-9013-2

Project and realization
Parramón, Inc.

Texts
Adolfo Cassan

Translator
Patrick Clark

Graphic Design and Typesetting
Toni Inglés Studio

Illustrations
Marcel Socías Studio

First edition - September 2004

Printed in Spain
© Parramón Ediciones, S.A. – 2005
Ronda de Sant Pere, 5, 4ª planta
08010 Barcelona (España)
Norma Editorial Group

www.parramon.com

TABLE OF CONTENTS

MARVELOUS FACULTIES

This book aims to give young readers basic information about the senses. The senses make up our connection to the outside world. They allow us to carry out our daily activities. They warn us of dangers. They bring us both pleasure and discomfort. It is interesting and valuable to learn about the different sensory organs and the way they work.

Our goal in creating this book was to make the subject practical, educational, challenging, and at the same time, entertaining for the reader.

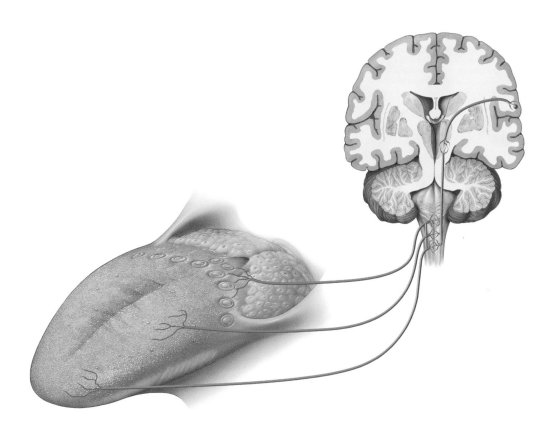

IN CONTACT WITH THE OUTSIDE WORLD

The information our brain receives from the different sense organs lets us know about the world around us.

taste ■

smell ■

■ touch

■ hearing

■ sight

The senses are special faculties that let us receive and recognize all kinds of information that comes from our surroundings and from inside our bodies. Human beings have five basic senses that let us interact with the world around us: sight, hearing, touch, smell, and taste. We might also add another sense—balance—which allows us to know where our body is located and its position at any given moment.

Our vision is nowhere near as sharp as that of an eagle, and our hearing ability is not as keen as a dog's, but our senses still give us the information we need to carry out the activities associated with daily life. Our senses tell us what is going on in the world around us. They tell us where something is located. They help us avoid dangers, and they let us recognize the people around us. Without our senses, we would be completely isolated in the world.

SENSORY RECEPTORS

All information that comes to us from the outside world arrives by means of physical or chemical stimuli: light rays, sound waves, and chemical particles in the air we breathe or in the food we eat. To register these stimuli, we need special receptors that can detect them, and we also need these receptors to turn the stimuli into another kind of signal that the brain can understand. The brain is something like our "central computer." It makes us aware of sensations.

Vision not only lets us appreciate the shape and size of things, but it also provides us with a wonderful, colorful world.

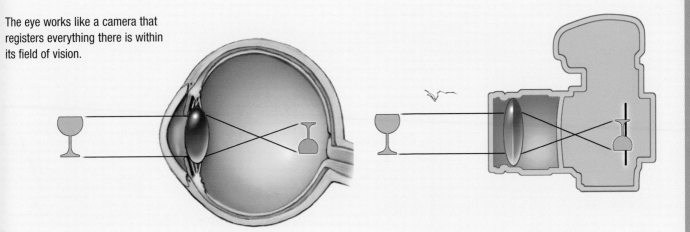

The eye works like a camera that registers everything there is within its field of vision.

Specific receptors capture stimuli that correspond to each of our senses: photoreceptors in the eye take note of light stimuli; cells in the ear detect sounds; many specialized cells on the surface of the skin feel even the slightest touch; receptors in the nose recognize fragrant particles in the air we breathe; and taste buds on the tongue note the flavor of everything we take into our mouths.

Sensory receptors are like microchips with a precise function. They take physical or chemical stimuli and turn them into electrical impulses, which are "interpreted" by our central nervous system. The sensory receptors start a complex process that allows our brains to recognize the stimuli. From the time stimuli are detected until we become fully aware of the sensations, the impulses generated in the receptors have a long way to travel.

THE PATH OF SENSATIONS

All stimuli that come from outside the body and are registered by sensory receptors have a common destination: the brain, the central computer of our body, where all information is gathered and interpreted. To get to the brain, stimuli must follow a long route along specific pathways of nerve cells.

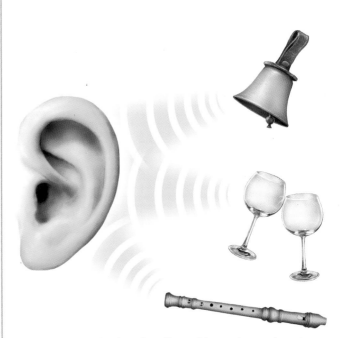

The ear takes in vibrations from the world around us, and sends information about them to the brain. The brain then decodes these messages and identifies sound sources with the greatest possible precision.

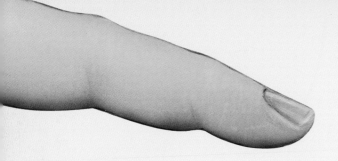

Each of our senses has a protective function, alerting us to potential danger before it can harm us.

In human beings, the sense of smell may not be as critical for survival as it is for many animals. Nevertheless, it does allow us to enjoy many pleasant aromas.

These cells, called neurons, connect the various sense organs with the brain. After receiving electrical impulses generated by the sense receptors, sensory nerves carry messages to the central nervous system (made up of the brain and spinal cord). Sometimes these messages travel only short distances, as in the case of the olfactory (smell-related) nerve, which only has to follow a path from the nose to the brain, or the optical nerve, which begins at the back of the eye. But at other times, the path is much longer: Just think about how far the signal produced by tickling the big toe has to travel to arrive at the head!

Along this route to the brain, electrical impulses travel over at least three clusters of sensory neurons before they reach the cerebral cortex, where they are processed by the brain. For the most part, these sensory pathways to the cerebral cortex are crossed pathways. This means that each side of the brain registers sensations from the opposite side of the body.

THE BRAIN: THE REAL SENSE ORGAN

It is on the surface of the brain, in the cerebral cortex, that sensations actually become part of our consciousness. In the cerebral cortex the stimuli that began as light rays become images, sound waves that penetrated the ear are turned into sounds or melodies, or we identify a smell, recognize a flavor, or feel a caress.

The brain has a great ability to interpret stimuli and give us an idea of what the world around us is like. It lets us recognize the face of a friend, a song by our favorite band, the scent of a rose, or the flavor of chocolate. Simple physical and chemical stimuli are converted into colorful images, harmonious melodies, exquisite fragrances, tickles, and caresses.

When it is born, a baby already has all its senses. The best developed of a baby's senses is the sense of touch. That is why babies love to be caressed.

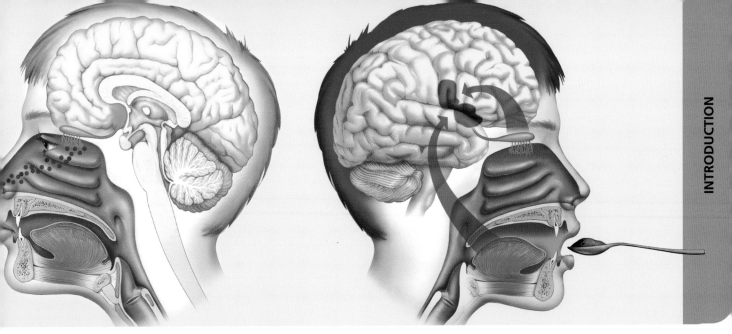

THE DEVELOPMENT OF THE SENSES

Not all of our senses are fully developed when we are born. Some of them still need to be perfected. It may seem strange, but the development of the different senses does not take place in an order that relates to how important each one of them is in our everyday lives.

The most fully developed sense in the newborn is touch. It is through the skin that a baby begins to know the world, can tell if it is being touched, and knows if it is dry, wet, cold, or hot. A baby's sense of taste is also well developed at birth—especially for sweets—which includes the flavor of milk, a baby's first food. It will take a baby close to two years to recognize all the flavors it tastes. And while a newborn reacts to strong odors, it barely responds to most milder smells, with one exception: the smell emitted by its own mother, which it recognizes without fail.

Taste and smell are closely related senses. In fact, the information both of them provide combine to give us our sense of taste.

In contrast, hearing and sight are poorly developed at birth. A baby barely responds to sounds, even intense ones, until about the age of 4 months, when it will turn its head in the direction from which a loud noise is coming. But it is not until the age of 8 months that the baby turns its head toward a familiar voice, and not until 18 months that it responds to sounds coming from far away. With regard to sight, a baby's sight is quite poor at birth. It is not until the baby is 2 months old that it begins to distinguish the profile of its mother's face. At the age of 3 months, it follows the movement of light across its field of vision. The baby begins to perceive colors in the following order: yellow first, then blue, red, and green. Even though a 6-month-old baby can distinguish color differences and is capable of focusing on specific objects, it is not until a few years later that its sense of sight reaches its full potential.

A WINDOW ON THE WORLD

The eye is a complex and delicate structure that receives light stimuli from outside the body and turns those stimuli into nerve impulses. These impulses are then sent through the optic nerve to the brain, where they are decoded and interpreted as images. The eye's function may be compared to that of a camera or, better yet, to a movie camera, since it gives a constant, moving, visual representation of the world around us.

conjunctiva ■
transparent membrane that covers the front of the eye and the internal face of the eyelids, protecting them from foreign substances

crystalline lens ■
transparent, elastic disk that focuses light rays on the surface of the retina

cornea ■
transparent disk that protects the front of the eye and lets light pass into the eyeball

LIGHT EYES, DARK EYES

The color of the eyes depends on how much of a pigment called melanin is found in the iris. Melanin is also responsible for skin color. The iris is blue when the amount of melanin is low, as in persons with very pale skin. The iris is darker depending on how much more pigment it contains. People who have dark skin also tend to have brown eyes.

iris ■
pigmented (colored) disk that takes in light rays and lets them pass into the eyeball through an orifice (hole) called the pupil located in its center

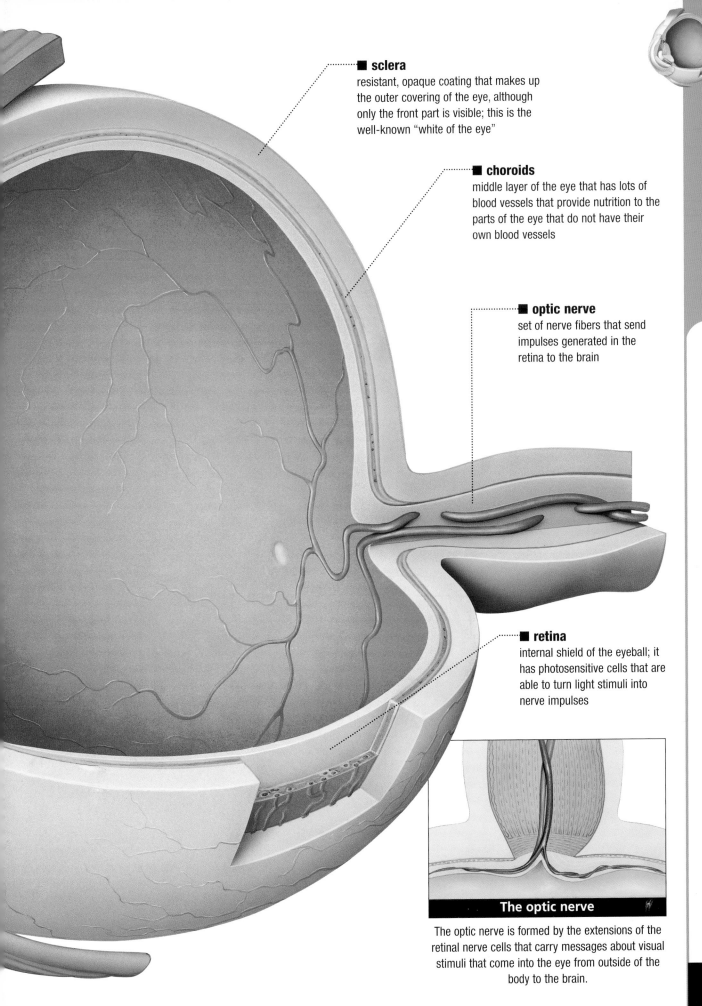

■ **sclera**
resistant, opaque coating that makes up the outer covering of the eye, although only the front part is visible; this is the well-known "white of the eye"

■ **choroids**
middle layer of the eye that has lots of blood vessels that provide nutrition to the parts of the eye that do not have their own blood vessels

■ **optic nerve**
set of nerve fibers that send impulses generated in the retina to the brain

■ **retina**
internal shield of the eyeball; it has photosensitive cells that are able to turn light stimuli into nerve impulses

The optic nerve

The optic nerve is formed by the extensions of the retinal nerve cells that carry messages about visual stimuli that come into the eye from outside of the body to the brain.

THE MIRACLE OF VISION

Visual perception of the world around us depends on a complex process called ocular refraction. Thanks to ocular refraction, the light rays coming from objects in our field of vision become focused on the surface of the retina. But something very strange happens in this process: The images of objects are focused on the retina in an inverted (upside-down) position. However, the brain automatically interprets them to be right-side up.

FROM NEAR OR FAR

To see an object well, we need to focus on it. If we focus on an object very close to us, then we cannot clearly see something that is far away. On the other hand, if we look into the distance, we can see the nearest objects in all their detail. Fortunately, changes of focus happen automatically: All we need to do is look at what we want to see for the eyes to focus the correct way.

2 cornea and lens ■
divert light rays to focus them on the retina

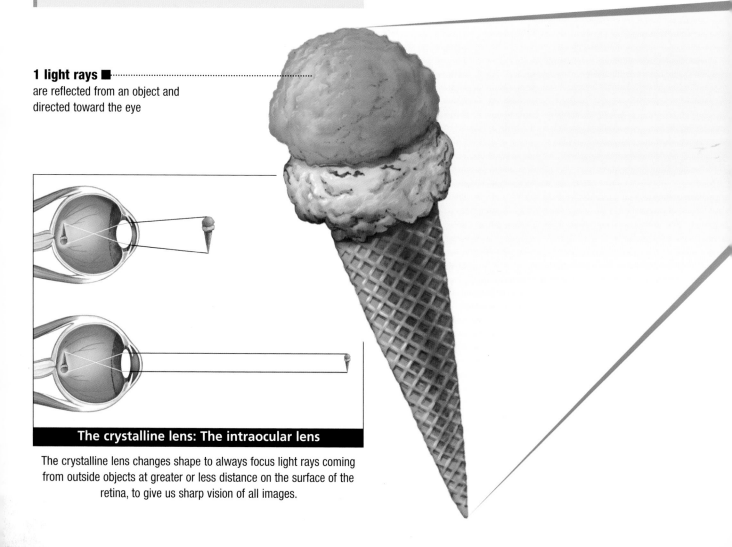

1 light rays ■
are reflected from an object and directed toward the eye

The crystalline lens: The intraocular lens

The crystalline lens changes shape to always focus light rays coming from outside objects at greater or less distance on the surface of the retina, to give us sharp vision of all images.

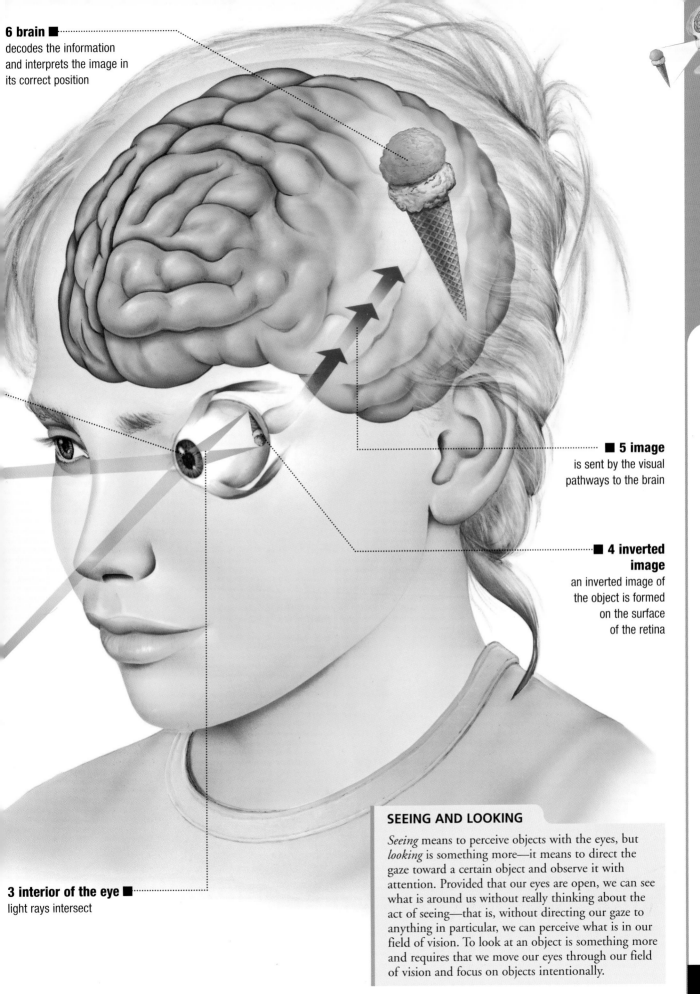

6 brain ■
decodes the information
and interprets the image in
its correct position

■ 5 image
is sent by the visual
pathways to the brain

**■ 4 inverted
image**
an inverted image of
the object is formed
on the surface
of the retina

3 interior of the eye ■
light rays intersect

SEEING AND LOOKING

Seeing means to perceive objects with the eyes, but
looking is something more—it means to direct the
gaze toward a certain object and observe it with
attention. Provided that our eyes are open, we can see
what is around us without really thinking about the
act of seeing—that is, without directing our gaze to
anything in particular, we can perceive what is in our
field of vision. To look at an object is something more
and requires that we move our eyes through our field
of vision and focus on objects intentionally.

THE WORLD IN COLOR

The retina of the eye contains photoreceptors—cells that react to light rays and create nerve impulses that the brain interprets as images. Some photoreceptors, called rods, are stimulated at night or when there is little light, and only allow vision in black and white. Other photoreceptors, called cones, are active by day or in well-lit places, and detect colors. The eye has three types of cones, each of which is sensitive to a primary color—blue, red, or green. Using all three types at once allows us to see the world in an infinite range of colors.

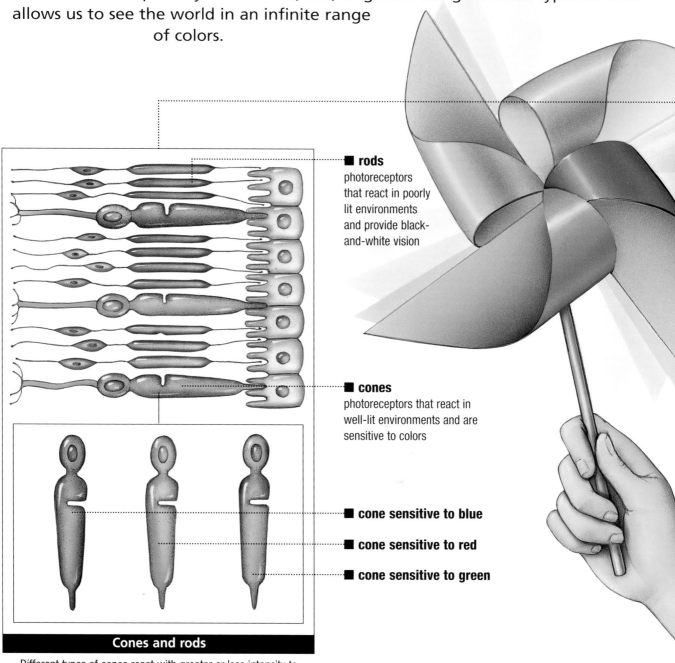

rods
photoreceptors that react in poorly lit environments and provide black-and-white vision

cones
photoreceptors that react in well-lit environments and are sensitive to colors

cone sensitive to blue

cone sensitive to red

cone sensitive to green

Cones and rods

Different types of cones react with greater or less intensity to different colors. Having each type partly stimulated creates many combinations that allow the brain to perceive hundreds of colors and color combinations.

COLOR BLINDNESS

Around 5% of males and 1% of females do not perceive or distinguish all colors well, especially red and green. They suffer a hereditary disorder known as color blindness. This condition is sometimes called Daltonism, named in honor of John Dalton, a prestigious British chemist and physicist who suffered from the condition and described it in great detail toward the end of the 18th century.

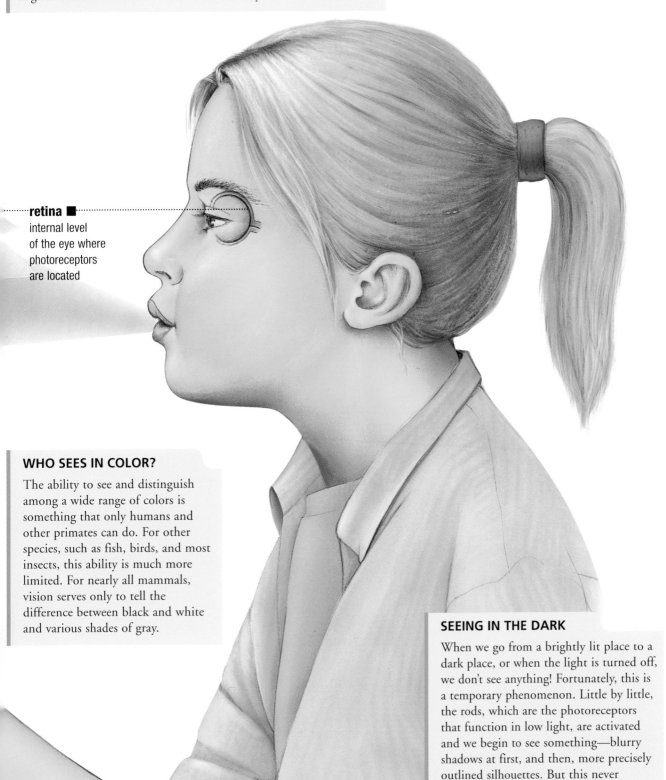

retina ■
internal level
of the eye where
photoreceptors
are located

WHO SEES IN COLOR?

The ability to see and distinguish among a wide range of colors is something that only humans and other primates can do. For other species, such as fish, birds, and most insects, this ability is much more limited. For nearly all mammals, vision serves only to tell the difference between black and white and various shades of gray.

SEEING IN THE DARK

When we go from a brightly lit place to a dark place, or when the light is turned off, we don't see anything! Fortunately, this is a temporary phenomenon. Little by little, the rods, which are the photoreceptors that function in low light, are activated and we begin to see something—blurry shadows at first, and then, more precisely outlined silhouettes. But this never happens immediately. Instead, a few minutes must pass for our eyes to get used to the darkness.

HOW IMAGES ARE FORMED

Light rays that hit the retina of the eye follow a long route to arrive at their destination: the visual area located in the occipital lobe of the cerebral cortex. In this area, nerve impulses are decoded, turned into visual sensations, and then these perceptions are interpreted by the brain. That is, the brain forms recognizable visual representations of objects.

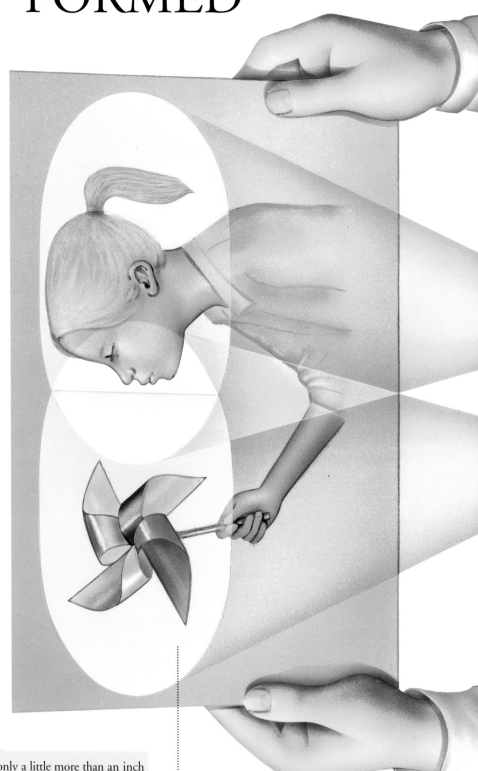

THREE-DIMENSIONAL VISION

Because both of our eyes, which are only a little more than an inch apart, see the same objects at the same time, the images that reflect on the retina of each eye are slightly different from the actual object that we see. Both eyes working at once give us stereoscopic vision that lets us see the contours of objects and judge depth.

■ **field of vision**

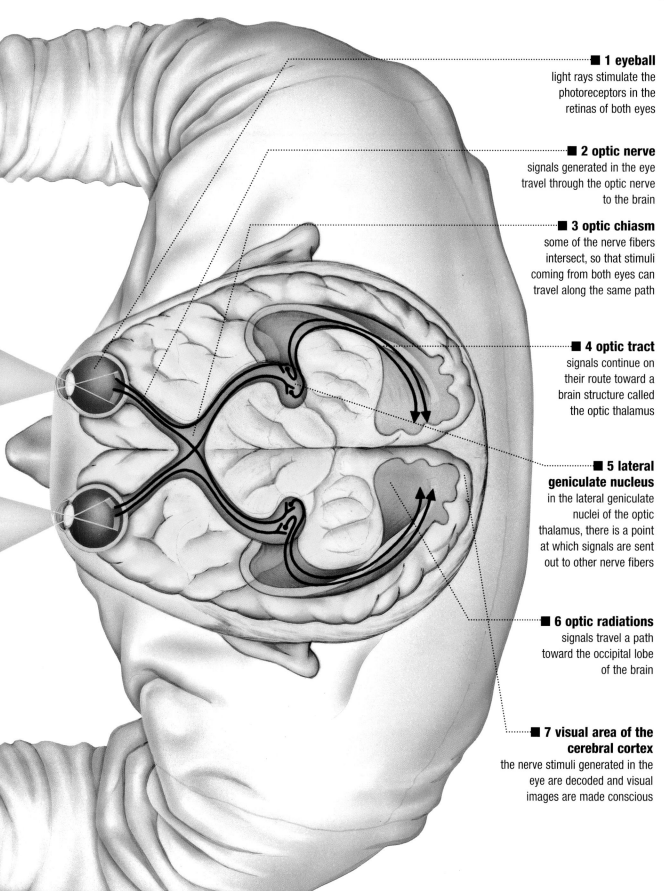

■ **1 eyeball**
light rays stimulate the photoreceptors in the retinas of both eyes

■ **2 optic nerve**
signals generated in the eye travel through the optic nerve to the brain

■ **3 optic chiasm**
some of the nerve fibers intersect, so that stimuli coming from both eyes can travel along the same path

■ **4 optic tract**
signals continue on their route toward a brain structure called the optic thalamus

■ **5 lateral geniculate nucleus**
in the lateral geniculate nuclei of the optic thalamus, there is a point at which signals are sent out to other nerve fibers

■ **6 optic radiations**
signals travel a path toward the occipital lobe of the brain

■ **7 visual area of the cerebral cortex**
the nerve stimuli generated in the eye are decoded and visual images are made conscious

CAPTURING SOUNDS

The ear is an extraordinary organ. We can only see the outermost portion of it, since the rest is located inside the head. Although everyone knows that the ear is responsible for hearing, we often fail to realize that the inner ear also has a role in another sense—the sense of balance—which allows us to stay on our feet or to move and turn without falling.

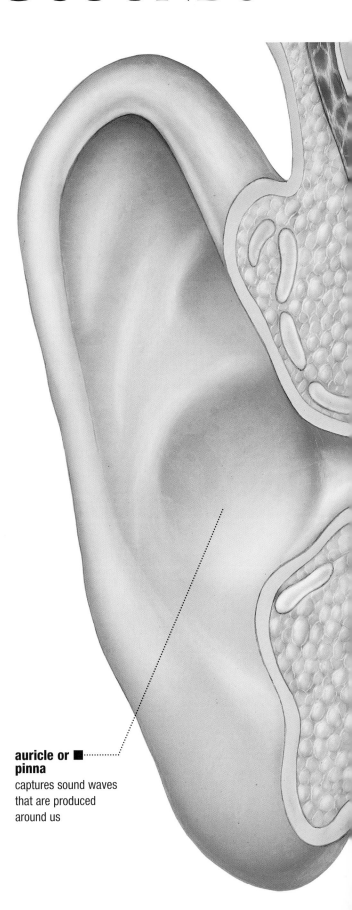

auricle or ■ ·········
pinna
captures sound waves
that are produced
around us

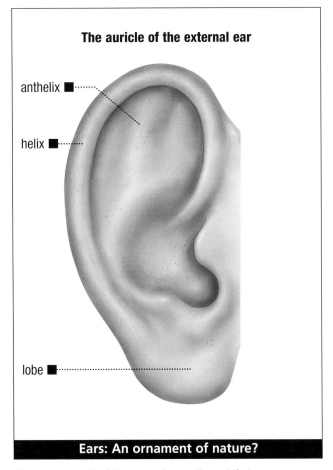

The auricle of the external ear

anthelix ■ ·········

helix ■ ·········

lobe ■ ·········

Ears: An ornament of nature?

In humans, much of the external ear—the part that we can see—has little effect on our ability to hear. Unlike animals, we cannot move our ears to focus on the direction of sounds that are within our range of hearing. In fact, our ability to hear would not be greatly reduced if we were missing much of the outer portion of our ears.

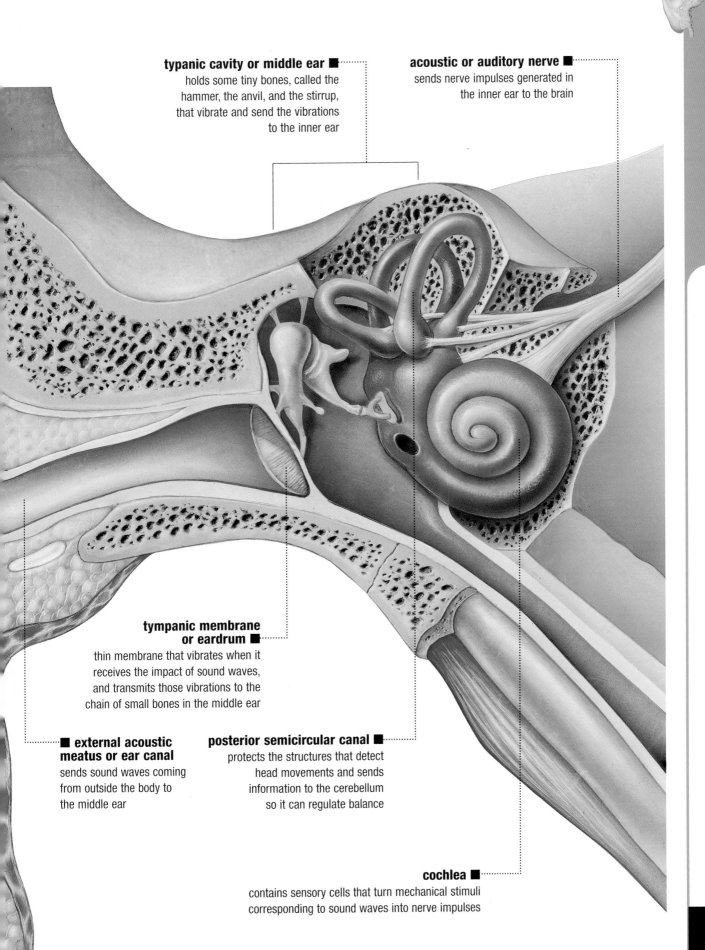

typanic cavity or middle ear ■
holds some tiny bones, called the
hammer, the anvil, and the stirrup,
that vibrate and send the vibrations
to the inner ear

acoustic or auditory nerve ■
sends nerve impulses generated in
the inner ear to the brain

**tympanic membrane
or eardrum ■**
thin membrane that vibrates when it
receives the impact of sound waves,
and transmits those vibrations to the
chain of small bones in the middle ear

**■ external acoustic
meatus or ear canal**
sends sound waves coming
from outside the body to
the middle ear

posterior semicircular canal ■
protects the structures that detect
head movements and sends
information to the cerebellum
so it can regulate balance

cochlea ■
contains sensory cells that turn mechanical stimuli
corresponding to sound waves into nerve impulses

HELLO . . . I HEAR YOU

Hearing is the sense that allows us to turn mechanical stimuli such as sound waves— vibrations of air molecules that expand from the place where a sound is produced—into nerve stimuli, which the brain interprets as sounds. In addition to helping us perceive what is happening around us, our sense of hearing is also important as a basic tool for communication and spoken language, the main way human beings relate to each other.

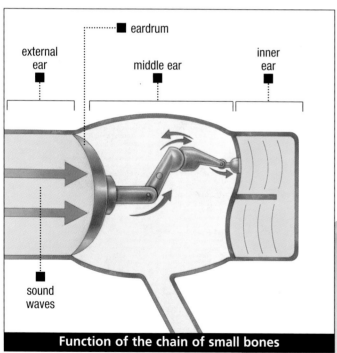

■ eardrum

external ear ■

middle ear ■

inner ear ■

sound waves

Function of the chain of small bones

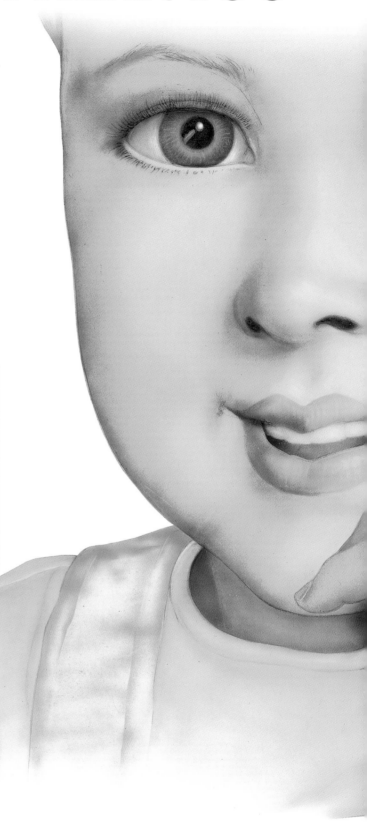

THE REAL ORGAN OF HEARING

Sensory cells located in the inner ear turn the mechanical energy of sound waves into the electrical energy of the signals that the auditory (hearing-related) nerves send to the brain. These cells make up the organ of Cortis, the real organ of hearing.

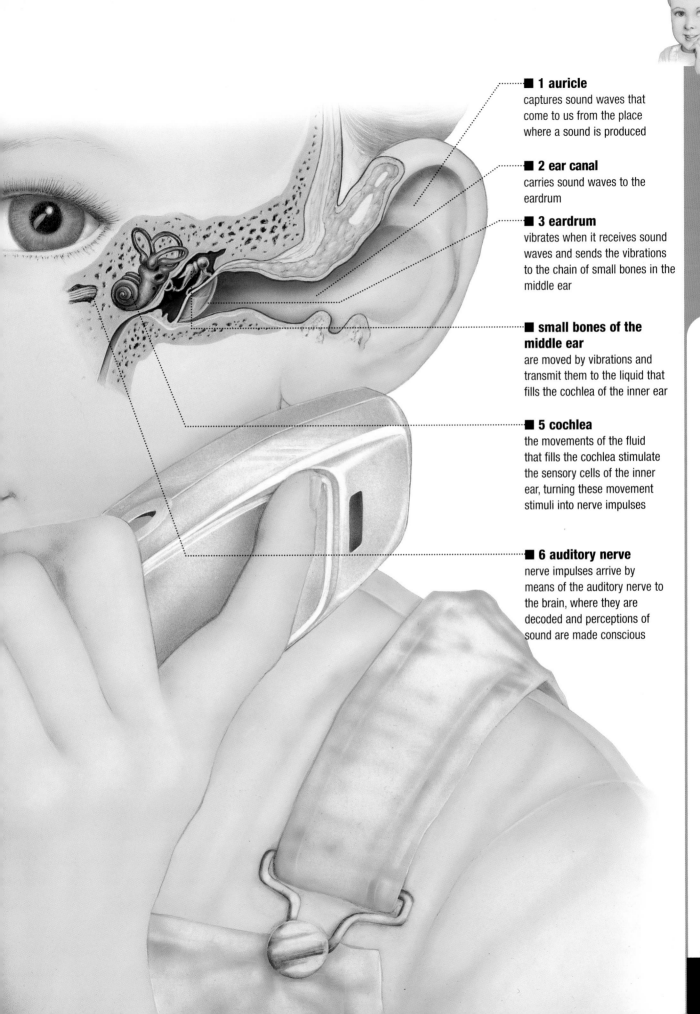

■ **1 auricle**
captures sound waves that come to us from the place where a sound is produced

■ **2 ear canal**
carries sound waves to the eardrum

■ **3 eardrum**
vibrates when it receives sound waves and sends the vibrations to the chain of small bones in the middle ear

■ **small bones of the middle ear**
are moved by vibrations and transmit them to the liquid that fills the cochlea of the inner ear

■ **5 cochlea**
the movements of the fluid that fills the cochlea stimulate the sensory cells of the inner ear, turning these movement stimuli into nerve impulses

■ **6 auditory nerve**
nerve impulses arrive by means of the auditory nerve to the brain, where they are decoded and perceptions of sound are made conscious

I'M NOT FALLING!

Unlike the five basic senses, which bring us information about the outside world, the sense of balance has another mission: to provide information to the brain about the position and movements of our body. The brain uses this information to adjust the tension of our muscles, which keep us from falling down. This function is extremely important for creatures like humans that, unlike most terrestrial (land) animals, move around by using only two extremities (our legs), instead of four.

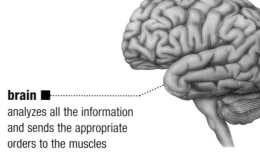

brain ■
analyzes all the information and sends the appropriate orders to the muscles

eyes ■
provide the brain with an overall idea of where the body is and give us external reference points

AROUND AND AROUND

It is very easy to demonstrate what happens when there is a problem with your sense of balance. Spin around rapidly several times with your eyes closed, then stop suddenly and open your eyes. It will seem to you that everything is spinning; your muscles will probably fail, and you will have a tendency to fall down. This is because the fluid that fills the semicircular canals of the inner ear will still be moving, and the brain will be a little slow to notice that the information it is receiving from the inner ear does not match reality—that you have stopped turning. But don't worry—the brain will soon realize what is happening and everything will return to normal.

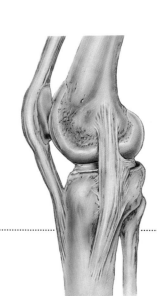

muscles ■
some contract and others relax, to maintain a position or make movements without succumbing to the force of gravity

receptors in the joints ■
provide information to the brain about the relative positions of the different parts of the body

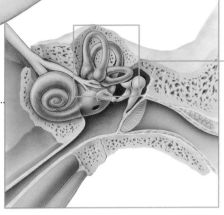

■ the crista (found in the semicircular canals)
tells the brain about the position and movements of the head

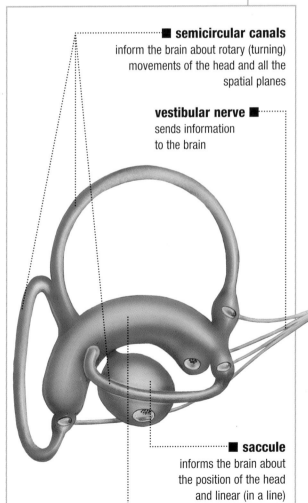

■ semicircular canals
inform the brain about rotary (turning) movements of the head and all the spatial planes

vestibular nerve ■
sends information to the brain

■ saccule
informs the brain about the position of the head and linear (in a line) movements in the vertical plane

utricle ■
provides information to the brain about the position of the head and linear movements on the horizontal plane

RECOGNIZING WHAT WE TOUCH

Touch is the sense that allows us to recognize the shape and size of objects, whether their surfaces are smooth and soft or rough and harsh, or whether their temperature is cold or hot. The organ where this sense resides is the skin—in particular, the skin of the hands, and especially the fingertips, inside of which there are different types of special receptors that can detect different types of stimuli.

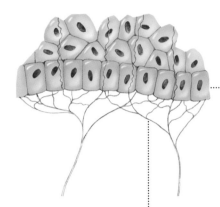

free nerve endings ■
perceive touch stimuli, but react mainly to painful stimuli

epidermis ■
top layer of the skin

dermis ■
middle layer of the skin

THE MOST PRECOCIOUS OF THE SENSES

After the 13th week of gestation (the period during which a fetus develops), the sensitivity receptors of the skin have already begun to develop. Touch will be the first sense with which a newborn baby begins to relate to its surroundings.

Meissner's corpuscles ■
very abundant in the fingertips and in the lips; respond to touch stimuli

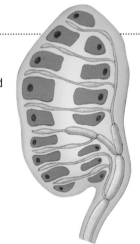

hypodermis ■
deep layer of the skin

TRAINING THE SENSE OF TOUCH

Touch is a sense that can be perfected with training. Doctors practice to be able to notice tiny differences when they touch the bodies of their patients. Touch is also a basic tool of sculptors and craftspeople, for technicians who work with tiny parts, and for the blind, who can partly make up for their loss of vision with touch, by learning to read with the Braille system, for example.

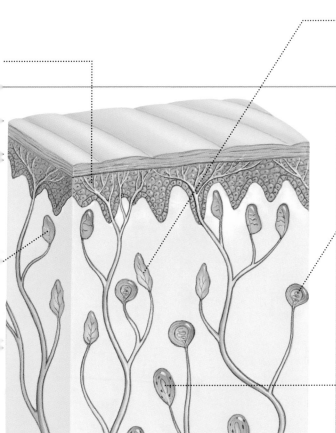

■ **Ruffini corpuscles**
detect thermal (temperature) stimuli, and are especially sensitive to heat

■ **Krause corpuscles**
detect thermal stimuli, and are especially sensitive to cold

■ **Pacinian corpuscles**
perceive pressure changes and vibrations that are produced on the skin

THAT SMELLS GOOD!

Smell is the sense that lets us detect odors. These odors may be the aromas of food that enhance our appetite or perfumes that bring us pleasure, but they may also be strong smells that are unpleasant to us. Foul odors often warn us of danger, like rotten food or toxic gas.

INTENSE SMELLS

When we are exposed to intense smells for a long period of time, whether they are agreeable or disagreeable, olfactory cells "get tired" and eventually stop reacting. For this reason, we get used to very strong smells, and after a while, we barely notice them.

2 olfactory cells ■
generate nerve impulses that are sent to the olfactory bulb

1 olfactory epithelium ■
the tiny flaps of the olfactory cells make contact with particles in the air we breathe

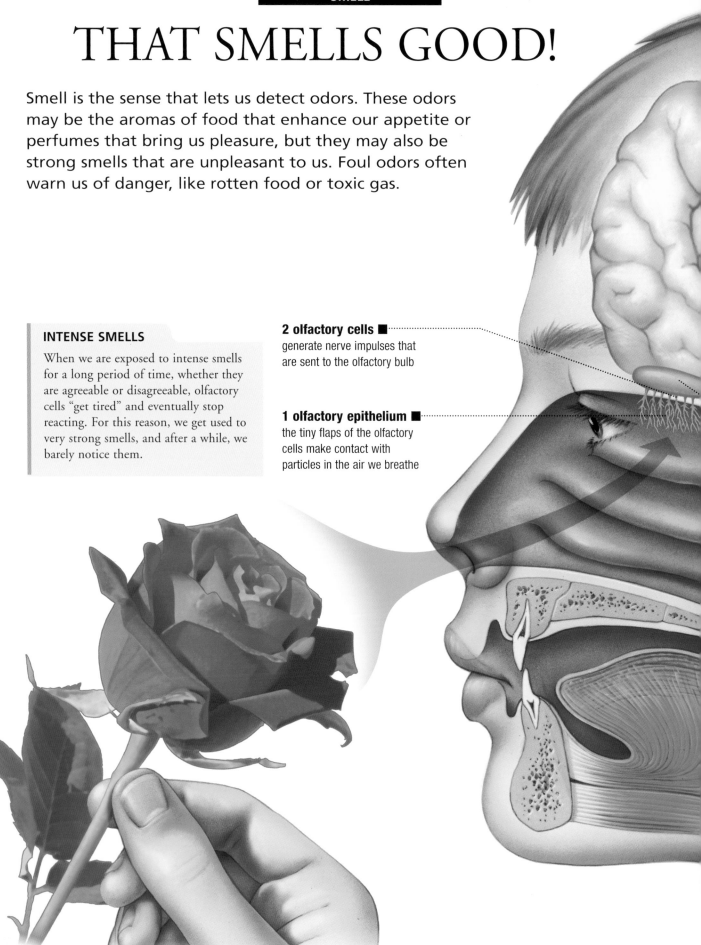

■ 5 olfactory areas of the brain

nerve signals are turned into conscious sensations

■ 4 olfactory nerve

sends signals to the olfactory areas of the brain

■ 3 olfactory bulb

the cells of the olfactory bulb are stimulated and produce signals that are transmitted by the olfactory nerve

SMELL AND MEMORY

Smell, more than the other senses, has a great power to bring back memories. A certain aroma may remind us of an experience from a vacation, our grandmother's house, or a day when something special happened. Although it may seem strange, it is likely that these smells will keep bringing up the same memories for us as years go by. Here's a suggestion: When something good happens to you, breathe deeply and try to capture the particular aroma of that moment. You can then keep these magical moments in your memory for your whole life!

SAVOR THE FLAVORS

Taste is a sense that allows us to recognize particular characteristics of foods, beverages, and everything we put in our mouths. On the surface of the tongue, there are thousands of tiny taste buds—the receptors for taste—that react with chemical substances that are dissolved in saliva and then send messages to the brain. These messages are converted into taste sensations, some delicious and others very unpleasant.

THE FUNCTION OF TASTE

Taste allows us to recognize what we are eating or drinking, even with our eyes closed! But it has a much more important function as well. When we taste a food we like, this encourages the body to produce digestive juices, helping us digest better and take advantage of the nutrition that the food contains.

MY MOUTH IS WATERING!

Chemical substances in foods stimulate the taste receptors only if they are dissolved in saliva. This is why our mouths water when we see a food we like.

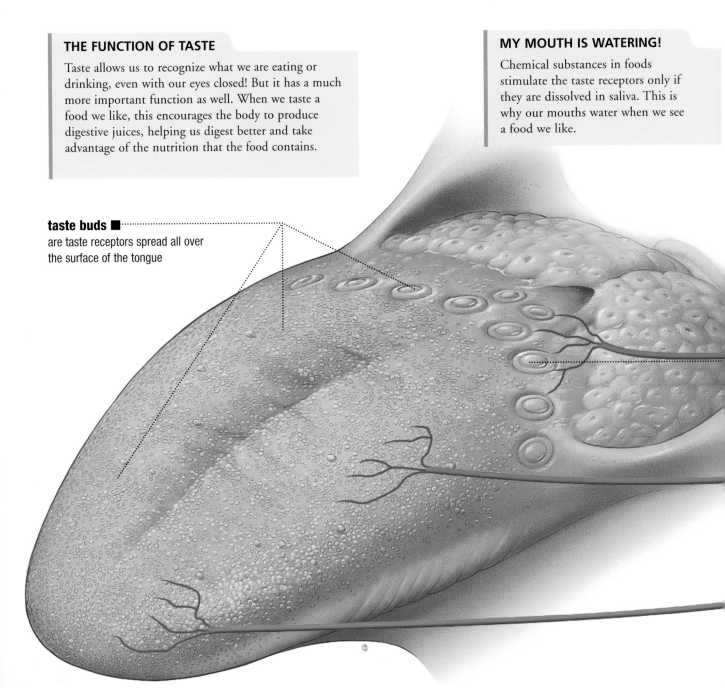

taste buds ■
are taste receptors spread all over the surface of the tongue

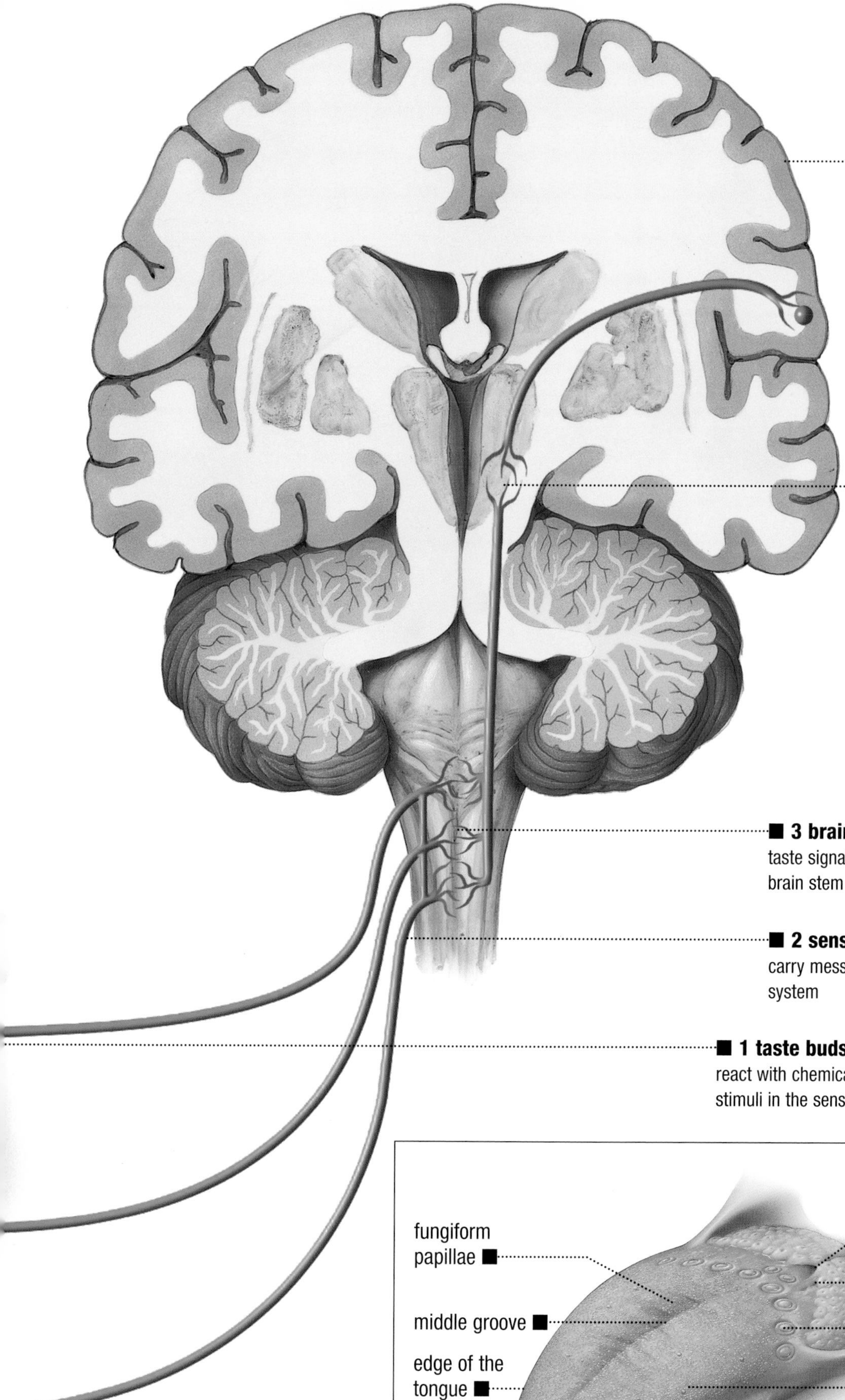

5 cerebral cortex
the signals arrive at the cerebral cortex, where taste sensations are made conscious

4 thalamus
taste stimuli make a second stop at the thalamus, a nerve center located inside the brain

3 brain stem
taste signals make a first stop at the brain stem

2 sensory nerves
carry messages to the central nervous system

1 taste buds
react with chemical substances to generate stimuli in the sensory nerves

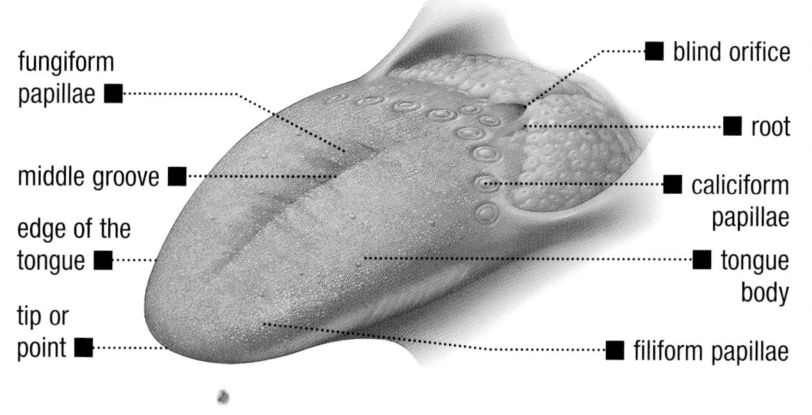

fungiform papillae ■

middle groove ■

edge of the tongue ■

tip or point ■

■ blind orifice

■ root

■ caliciform papillae

■ tongue body

■ filiform papillae

Parts of the tongue

HOW SWEET IT IS!

Our taste buds recognize only five basic sensations: sweet, sour, acid, salty, and a recently discovered taste called "savory" or *umami*, from the Japanese word for "delicious." This is a taste found in beef, cheeses, and some other foods that does not match any of the other four taste sensations. The brain combines the different tastes with the stimuli coming from the sense of smell and allows us to enjoy thousands of different tastes. Although taste buds are found all over the tongue, each of the basic taste sensations is perceived most intensely in specific areas of the tongue.

PERSONAL PREFERENCES

Almost everyone enjoys sweets, others prefer salty foods, few of us like acids, and very few enjoy bitter tastes. However, the sense of taste can be "educated": There are foods that do not seem tasty to us the first time we try them, but over time, they may become some of our favorites. It is a good idea to be bold and "learn" to enjoy the many combinations of tastes that nature offers us.

bitter ■
the bitter perception area is located at the back part of the tongue

salty ■
the salty perception area is located in the front part of the tongue, except for the tip

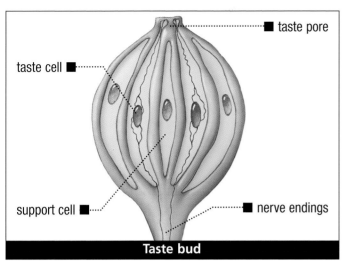

taste pore

taste cell ■

support cell ■ .... ■ nerve endings

Taste bud

Taste receptors are tiny pouch-like structures found on the surface of the tongue. They have various sensory cells and other support cells around a central cavity, into which chemical substances dissolved in saliva enter. These receptors are clustered in taste buds. When the sensory cells are stimulated, they generate impulses that travel through nerve endings to the brain.

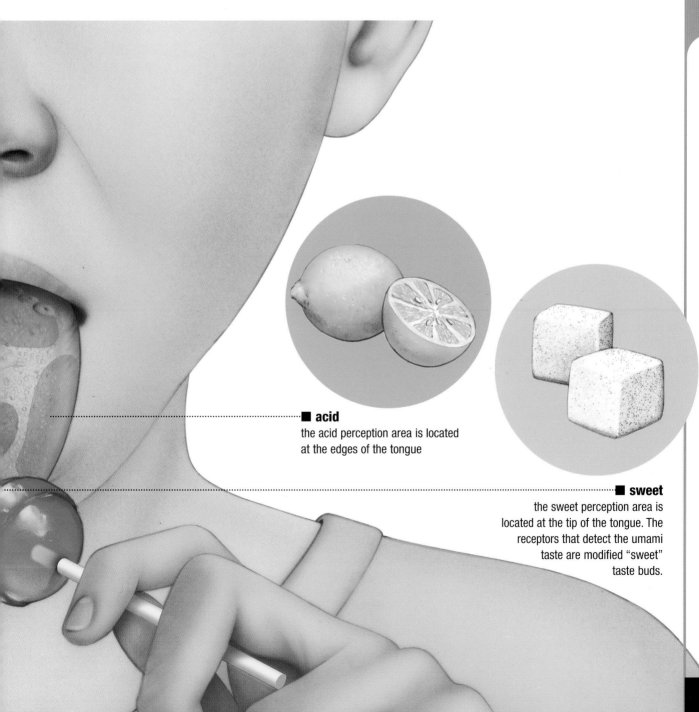

■ **acid**
the acid perception area is located at the edges of the tongue

■ **sweet**
the sweet perception area is located at the tip of the tongue. The receptors that detect the umami taste are modified "sweet" taste buds.

DID YOU KNOW?

HOW WELL CAN I TELL THE DIFFERENCES BETWEEN COLORS?

Look carefully at these drawings. If your color vision is good, you will be able to distinguish the following symbols, from left to right and top to bottom: 182, 13, F4, and 69.

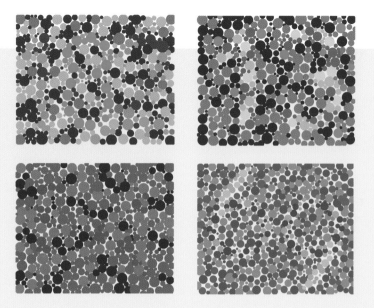

DO WE HEAR IT ALL?

Human hearing can only capture sound waves within a frequency range of 16 to 20,000 Hertz (vibrations per second). However, there are animals that are able to perceive sound waves that we cannot hear: those that have frequencies less than 20 Hertz (infrasounds), and those with frequencies over 20,000 Hertz (ultrasounds). This is why you can call a dog using an ultrasound whistle that the animal clearly hears, while we can't notice it at all.

For some animals, the perception of sound waves provides information essential for their survival. A bat can fly in the dark because it emits special sounds that bounce off objects and produce an echo that, coming back to the bat, allows it to tell when obstacles are present. We cannot hear these sound waves, because they are ultrasound waves. For the bat, however, these echoes enable it to "see" in the dark.

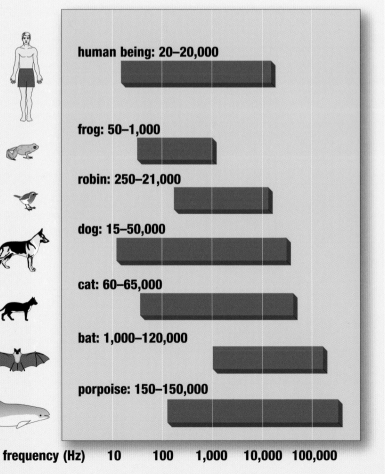

human being: 20–20,000

frog: 50–1,000

robin: 250–21,000

dog: 15–50,000

cat: 60–65,000

bat: 1,000–120,000

porpoise: 150–150,000

frequency (Hz) 10 100 1,000 10,000 100,000

INTERESTING FACTS

Measurements of the eye	The ocular globe has a spherical shape, although it is somewhat flattened in the vertical direction. Its diameter from the front part to the back, in an adult, is nearly one inch. Just a tiny fraction—hundredths of an inch—bigger or smaller creates a visual defect that requires a person to wear glasses or contact lenses.
The special tissue of the cornea	The cornea, which covers the front of the eye, needs to be transparent, because otherwise, light rays would not be able to get into the ocular globe. The transparency of the cornea is so important that its water content is over 75%. It does not even have its own blood irrigation system because blood vessels would get in the way of vision.
The diaphragm of the eye	The eye can regulate the amount of light rays that get inside it thanks to the control of the pupil, a small hole in the iris, the colored disk of the eyeball. In a poorly lit environment, the pupil will tend to remain more dilated (open), and will contract (get smaller) when a person passes into a room that is bright.
The tiny yellow spot	On the surface of the retina, there is an area of maximum visual acuity (sharpness), toward which light rays coming from inside are focused: the yellow spot, or macula lutea, is a tiny area of barely 5 square millimeters.
Looking at the point of the nose	To be able to see well, both eyes must focus on the same object. Otherwise, you would see double. This is what happens when you try to look at something that is too close to you. Try this: Put a finger 12 inches (31 cm) from your face and focus your glance on it. Without stopping to look at it, move your finger slowly toward you until it touches your nose. What happens?
Cones and rods	In the retina, there are around 130 million photoreceptors. Some 6 or 7 million are cones, responsible for color vision, and the rest are rods, responsible for black-and-white vision when we are in poorly lit environments.
The movements of the eye	We can direct our glance where we want thanks to six muscles found around the outer surface of each eyeball. These muscles act in a symmetrical manner. When we move our glance in one direction, the muscles located closest to that direction in each eye contract, and, at the same time, the muscles farthest away relax.
Let's blink!	It is estimated that we blink, on average, about 20,000 times a day. This is very important because the eyelids, with their continuous sweeping, keep the surface of the eye clean and moist with tear fluid.
Smell, a complementary sense	For a human being, smell is not as vital a sense as it is for many other animals, who are much more sensitive to odors, and depend on their sense of smell for survival.
Taste buds	We have about 10,000 taste buds spread out over the tongue, although there are also some taste buds in the palate and in the throat. Each papilla contains from 50 to 100 receptor cells that react with greater or less intensity to different stimuli. Thus, there are thousands and thousands of receptors that allow us to tell the difference between all kinds of tastes.

INDEX